HOW TO USE DIGITAL MULTIMETER

Everything You Need to Know About DMM Functions for Your Industrial and Domestic Usage

Richard Michael

Copyrigt@2024

Table of Contents

CHAPTER ONE 4
   Introduction 4
CHAPTER TWO 6
   Definition and Basic Functionality of Digital Multimeters 6
CHAPTER THREE 10
   Evolution and Advancements in Digital Multimeter Technology 10
CHAPTER FOUR 12
   Operating Principles 12
CHAPTER FIVE 19
   Basic Operations of a Digital Multimeter 19
CHAPTER SIX 26
   Voltage Measurement 26
CHAPER SEVEN 33
   Current Measurement 33
CHAPTER EIGHT 40
   Resistance Measurement 40
CHAPTER NINE 48
   Continuity and Diode Testing 48
CHAPTER TEN 54
   Advanced Functions of Digital Multimeters 54

CHAPTER ELEVEN 61

Data Logging and Recording 61

CHAPTER TWELVE 68

Importance in Electrical Work 68

CHAPTER THIRTEEN 75

Calibration and Maintenance of Digital Multimeters
.... 76

CHAPTER FOURTEEN 83

Safety Precautions and Best Practices for Using a
Digital Multimeter 83

CHAPTER ONE

Introduction

A digital multimeter (DMM) is an electronic instrument used to measure various electrical parameters such as voltage, current, and resistance. It typically consists of a display screen, a selection knob, input jacks for test leads, and internal circuitry for measurement.

Unlike analog multimeters, which use a moving needle to display readings, digital multimeters provide measurements in numerical form on a digital display.

Digital multimeters (DMMs) are essential tools in the field of electrical engineering, electronics, and various other industries where precise measurements of electrical parameters are required.

These instruments have replaced analog multimeters due to their enhanced accuracy, reliability, and additional features.

CHAPTER TWO

Definition and Basic Functionality of Digital Multimeters

A digital multimeter is a handheld electronic instrument used to measure multiple electrical parameters within a circuit or electrical system. These parameters typically include voltage, current, resistance, capacitance, frequency, and temperature.

Unlike analog multimeters, which use a moving needle to display readings on a scale, digital multimeters provide numerical measurements on a digital display. This digital readout offers greater precision and readability compared to analog counterparts.

The basic functionality of a digital multimeter involves selecting the desired measurement function using a rotary switch or selector knob, connecting the test leads to the appropriate input jacks, and observing the measured value displayed on the digital screen. Depending on the model, digital multimeters may offer various measurement ranges, modes, and additional features such as data logging, peak hold, and auto-ranging capabilities.

Comparison with Analog Multimeters.

Analog multimeters, also known as analogue multimeters, were widely used before the advent of digital technology. These instruments utilize a moving pointer or needle to indicate measured values on a scale printed on the instrument's face. While analog multimeters are simple to use and do not require batteries for operation, they have several limitations compared to digital multimeters.

One significant advantage of digital multimeters over analog counterparts is their higher accuracy and resolution. Digital displays provide precise numerical readings, making it easier to interpret measurements, especially for small changes in values.

Additionally, digital multimeters offer faster response times and are less susceptible to parallax errors associated with analog needle movements.

Another advantage of digital multimeters is their ability to measure a wider range of electrical parameters and provide additional functionalities such as capacitance, frequency, and temperature measurement. Digital multimeters also offer features like auto-ranging, which automatically selects the appropriate measurement range, simplifying the measurement process for users.

CHAPTER THREE

Evolution and Advancements in Digital Multimeter Technology

Since their inception, digital multimeters have undergone significant advancements and improvements in terms of technology, design, and functionality. Early digital multimeters had limited capabilities and were relatively bulky and expensive. However, advancements in integrated circuit technology, microprocessors, and display technologies have led to the development of more compact, affordable, and feature-rich digital multimeters.

Modern digital multimeters incorporate advanced features such as True RMS (Root Mean Square) measurement, which accurately measures AC signals with complex waveforms. They also offer enhanced safety features, including overload protection and fused input terminals to prevent damage from excessive voltage or current. Furthermore, digital multimeters may have connectivity options such as USB or Bluetooth for data transfer and remote monitoring.

Overall, digital multimeters have revolutionized the field of electrical measurements, offering unparalleled accuracy, versatility, and convenience. Their widespread adoption across various industries underscores their importance as indispensable tools for professionals and enthusiasts alike.

CHAPTER FOUR

Operating Principles

Explanation of how digital multimeters work.

Overview of internal components and circuitry,

Principles behind voltage, current, and resistance measurement.

Operating Principles of Digital Multimeters

Understanding the operating principles of digital multimeters (DMMs) is essential for using these instruments effectively in electrical measurements. Stated explanation covering how digital multimeters work, the internal components and circuitry, and the principles behind voltage, current, and resistance measurement.

Digital multimeters operate based on the principles of analog-to-digital conversion. When measuring electrical parameters such as voltage, current, or resistance, the DMM first converts the analog signal into a digital format using an analog-to-digital converter (ADC). This conversion process involves sampling the analog signal at regular intervals and quantizing each sample into a digital value.

Once the signal is digitized, it is processed by a microcontroller or digital signal processor (DSP) within the multimeter. The microcontroller performs necessary calculations and adjustments based on the selected measurement function and range.

Finally, the processed digital data is displayed on the multimeter's screen in numerical form.

Overview of Internal Components and Circuitry

Digital multimeters consist of several key internal components and circuitry that enable accurate measurement and display of electrical parameters.

Input Stage: This stage includes input jacks where test leads are connected to the DMM. Depending on the measurement function selected, the input stage may contain voltage dividers, current shunts, or resistance networks to properly scale and condition the input signal.

Analog-to-Digital Converter (ADC): The ADC is responsible for converting the analog input signal into digital data. It samples the input signal at a high frequency and assigns a digital value to each sample.

Microcontroller/Digital Signal Processor (DSP): The microcontroller or DSP processes the digitized data, performs necessary calculations (such as scaling and unit conversions), and controls the operation of the multimeter.

Display Module: The display module presents the measurement results in numerical form on a digital screen.

Modern digital multimeters often utilize liquid crystal displays (LCDs) or organic light-emitting diode (OLED) displays for improved visibility and energy efficiency.

Selector Switches and Control Buttons: These components allow users to select the desired measurement function, range, and additional settings such as hold, backlight, and auto-ranging.

Principles Behind Voltage, Current, and Resistance Measurement

Voltage Measurement: When measuring voltage, the DMM applies a known resistance (typically very high) across the circuit being tested and measures the resulting current flow. Using Ohm's Law ($V = IR$), where V is voltage, I is current, and R is resistance, the DMM calculates the voltage based on the measured current and known resistance.

Current Measurement: For current measurement, the DMM places a low-resistance shunt in series with the circuit being tested. The voltage drop across this shunt is proportional to the current flowing through the circuit.

Measuring this voltage drop and knowing the resistance of the shunt, the DMM calculates the current using Ohm's Law.

Resistance Measurement: To measure resistance, the DMM applies a known voltage across the resistor under test and measures the resulting current flow. Based on Ohm's Law, the resistance can be calculated using the ratio of voltage to current.

Understanding these principles is fundamental for accurate and reliable measurements using a digital multimeter. It enables users to interpret measurement results effectively and troubleshoot electrical circuits with confidence.

CHAPTER FIVE

Basic Operations of a Digital Multimeter

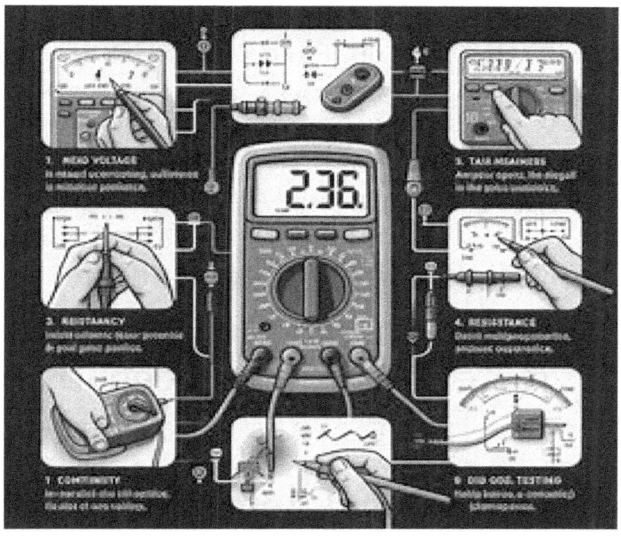

Mastering the basic operations of a digital multimeter (DMM) is crucial for accurate and efficient electrical measurements. Stated are explanations covering powering on/off the multimeter, selection of measurement functions, and choosing appropriate measurement ranges.

Powering On/Off the Multimeter

Powering On: To power on a digital multimeter, locate the power button or switch usually located on the front or side of the device. Press or toggle the power button to turn on the multimeter.

Upon powering on, the display screen may illuminate, and the multimeter may emit a beep to indicate that it's ready for use.

Powering Off: To power off the multimeter, simply press or toggle the power button again. Some digital multimeters may have an auto-power-off feature that automatically shuts down the device after a period of inactivity to conserve battery life.

Ensure the multimeter is powered off when not in use to prevent unnecessary battery drain.

Selection of Measurement Functions

Rotary Selector Dial: Digital multimeters typically feature a rotary selector dial or knob used to choose the desired measurement function.

Common measurement functions include voltage (DC and AC), current (DC and AC), resistance, capacitance, frequency, continuity, and diode testing.

Function Symbols: The selector dial is labeled with symbols or abbreviations representing each measurement function. Rotate the dial to align the pointer with the symbol corresponding to the parameter you intend to measure.

For example, select "V" for voltage measurement, "A" for current measurement, and "Ω" for resistance measurement.

Special Functions: Some digital multimeters may offer additional measurement functions or modes accessible through the selector dial. These functions could include temperature measurement, duty cycle measurement, and transistor testing. Consult the multimeter's user manual for instructions on accessing and using special functions.

Choosing Appropriate Measurement Ranges

Manual Range Selection: Digital multimeters may offer manual range selection or auto-ranging capabilities. With manual range selection, the user manually sets the measurement range using the selector dial. Each measurement function typically has multiple range options (e.g., 200mV, 2V, 20V, 200V, 1000V for voltage measurement).

Auto-Ranging: In auto-ranging mode, the multimeter automatically selects the most appropriate measurement range based on the magnitude of the measured signal. This feature simplifies the measurement process, as the user does not need to manually adjust the range.

However, auto-ranging may take slightly longer to stabilize compared to manual range selection.

Overload Protection: When selecting a measurement range, ensure that it is higher than the expected value of the measured parameter to avoid damage to the multimeter. Most digital multimeters are equipped with overload protection mechanisms that prevent damage from excessive voltage or current. If the measurement exceeds the selected range, the multimeter may display an error message or "OL" (overload) indication.

By mastering these basic operations, users can effectively utilize a digital multimeter to perform various electrical measurements with precision and confidence. It's essential to familiarize oneself with the specific features and functionalities of the multimeter through the user manual and practice using it in different measurement scenarios.

CHAPTER SIX

Voltage Measurement

Procedures for measuring DC and AC voltages.

Safety precautions while measuring voltage.

Accurate voltage measurement is a fundamental aspect of electrical diagnostics and troubleshooting.

Covering procedures for measuring both DC and AC voltages, safety precautions to observe, and how to interpret voltage readings.

Procedures for Measuring DC and AC Voltages

Measuring DC Voltage:

Set the digital multimeter to the DC voltage measurement function using the selector dial.

Connect the black (negative) test lead to the COM (common) terminal and the red (positive) test lead to the VΩmA (voltage, ohms, milliamp) terminal on the multimeter.

Ensure the multimeter is set to the appropriate DC voltage range higher than the expected voltage to be measured.

Place the test leads across the circuit or component being tested, ensuring correct polarity.

Read the voltage displayed on the multimeter's screen. Take note of any decimal point or units displayed.

Measuring AC Voltage:

Set the digital multimeter to the AC voltage measurement function using the selector dial.

Connect the test leads to the COM and VΩmA terminals as described for DC voltage measurement.

Choose the appropriate AC voltage range higher than the expected voltage to be measured.

Place the test leads across the circuit or component under test.

Read the AC voltage value displayed on the multimeter's screen.

Note the units (typically volts) and any decimal points.

Safety Precautions While Measuring Voltage

Isolate Power Sources: Before measuring voltage, ensure that the circuit or device under test is isolated from any power sources to prevent electric shock or damage to the multimeter.

Use Proper PPE: Wear appropriate personal protective equipment (PPE) such as insulated gloves and safety goggles when working with electrical systems to protect against potential hazards.

Inspect Test Leads: Check the test leads for any signs of damage or wear before use. Damaged test leads can compromise measurement accuracy and pose safety risks.

Avoid Contact with Live Circuits: Do not touch exposed conductors or components while voltage measurements are being taken, especially in live circuits.

Observe Polarity: Ensure correct polarity when connecting the test leads to the circuit or device under test to obtain accurate voltage readings.

Beware of High Voltages: Exercise caution when working with high-voltage systems. Use appropriate voltage-rated test leads and multimeters designed for high-voltage applications.

Interpretation of Voltage Readings

DC Voltage Readings: A DC voltage reading indicates the potential difference between two points in a circuit. Positive voltage readings indicate a higher potential at the red test lead compared to the black test lead, while negative readings indicate the opposite polarity.

AC Voltage Readings: AC voltage readings represent the magnitude and polarity of alternating voltage signals. The readings fluctuate between positive and negative values as the AC waveform alternates.

Stable Readings: Stable voltage readings indicate a steady voltage level in the circuit or device under test.

Fluctuating or erratic readings may indicate voltage instability or interference.

Comparative Analysis: Compare measured voltage readings to expected values or reference standards to assess the health and performance of electrical systems or components.

Understanding how to safely measure voltage and interpret voltage readings is essential for maintaining electrical safety and diagnosing potential issues within circuits or devices. Always follow proper procedures and precautions when working with electrical systems to mitigate risks and ensure accurate measurements.

CHAPER SEVEN

Current Measurement

Accurate current measurement is vital for assessing the flow of electrical charge within a circuit. Covering techniques for measuring both DC and AC currents, series and parallel connections for current measurement, and safety considerations for handling high-current measurements.

Techniques for Measuring DC and AC Currents

Measuring DC Current:

Set the digital multimeter to the DC current measurement function using the selector dial.

Connect the black (negative) test lead to the COM (common) terminal and the red (positive) test lead to the terminal marked for current measurement, typically labeled as "A" or "mA."

Ensure the multimeter is set to the appropriate DC current range higher than the expected current to be measured.

Break the circuit at the point where current measurement is required and insert the multimeter in series with the circuit, ensuring correct polarity.

Read the current value displayed on the multimeter's screen. Note the units (typically amperes or milliamperes) and any decimal points.

Measuring AC Current

Some digital multimeters have the capability to measure AC current directly, while others require the use of an additional current clamp accessory.

If the multimeter has a built-in AC current measurement function, follow similar steps as for DC current measurement, but ensure the multimeter is set to the AC current mode.

If using a current clamp accessory, clamp the jaws of the current clamp around a single conductor carrying the AC current. Follow the instructions provided with the current clamp for proper usage.

Series and Parallel Connections for Current Measurement

Series Connection:

When measuring current in a series circuit, the digital multimeter is connected in series with the circuit component being tested. This means that the current flows through the multimeter, allowing it to measure the current passing through the circuit.

Ensure the multimeter is inserted in series with the circuit by breaking the circuit and connecting the multimeter's test leads in-line with the current path.

Parallel Connection:

In some scenarios, it may be necessary to measure current in a parallel branch of a circuit.

However, directly measuring current in parallel branches is not typically feasible with a digital multimeter.

Instead, use the principles of Kirchhoff's current law to calculate the total current flowing into a junction or node in the circuit by measuring currents in series branches that converge at that point.

Handling High-Current Measurements Safely

Use Proper Equipment: When measuring high currents, ensure that the digital multimeter and test leads are rated for the expected current levels to prevent damage and ensure accurate measurements.

Avoid Overloading: Select the appropriate current range on the multimeter to avoid overloading the instrument. Start with the highest range and work your way down if necessary to obtain a more precise measurement.

Use Current Clamps: For extremely high currents, consider using a current clamp accessory designed for measuring large currents. This allows for non-intrusive measurements without breaking the circuit.

Safety Precautions: Always adhere to safety precautions when working with high currents, including wearing appropriate PPE such as insulated gloves and safety goggles. Be cautious of potential hazards such as heat dissipation and magnetic fields associated with high-current circuits.

By understanding these techniques and safety considerations, users can effectively measure current in both DC and AC circuits while ensuring their safety and the integrity of the measurement equipment.

CHAPTER EIGHT

Resistance Measurement

Methods for measuring resistance.

Techniques for accurate resistance readings.

Resistance measurement is essential for diagnosing faults, verifying component values, and analyzing circuits. Explanation covering methods for measuring resistance, techniques for accurate readings, and applications in circuit troubleshooting.

Methods for Measuring Resistance

Direct Measurement:

Set the digital multimeter to the resistance (Ω) measurement function using the selector dial.

Ensure the circuit or component under test is disconnected from any power source to prevent inaccurate readings.

Connect the black (negative) test lead to the COM (common) terminal and the red (positive) test lead to the terminal marked for resistance measurement.

Place the test leads across the resistor or component being tested. Ensure a good connection and avoid touching the test leads during measurement.

Read the resistance value displayed on the multimeter's screen. Note the units (typically ohms) and any decimal points.

Four-Point Probe Method:

This method is used for accurately measuring very low resistances, typically in the milliohm range.

It involves passing a known current through the resistor using two probes and measuring the voltage drop across the resistor using another two probes.

The resistance is calculated using Ohm's Law ($R = V/I$), where R is resistance, V is voltage, and I is current.

Techniques for Accurate Resistance Readings:

Zero Out Test Lead Resistance

Before taking resistance measurements, short the test leads together and press the relative (REL) button on the multimeter to zero out the resistance of the test leads. This compensates for any resistance introduced by the test leads themselves.

Stable Connection:

Ensure a stable and secure connection between the test leads and the component under test. Any loose or intermittent connections can result in inaccurate readings.

Minimize Contact Resistance:

Avoid touching the test leads or the component terminals during measurement to minimize contact resistance. Contact resistance can introduce additional resistance into the circuit, leading to inaccuracies in the measurement.

Temperature Compensation:

Be aware of temperature effects on resistance measurements, especially when measuring components with temperature-dependent resistances such as thermistors. Consider compensating for temperature variations to obtain accurate readings.

Applications of Resistance Measurement in Circuit Troubleshooting:

Component Testing

Resistance measurement is used to test the integrity of resistors, capacitors, and other passive components in a circuit. Abnormal resistance values may indicate component failure or out-of-spec conditions.

Short and Open Circuit Detection:

Resistance measurement helps identify short circuits (low resistance) and open circuits (infinite resistance) within a circuit. By measuring resistance at various points in the circuit, faults can be pinpointed for troubleshooting.

Continuity Testing:

Resistance measurement is used for continuity testing to verify the presence of a continuous electrical path between two points in a circuit. A low resistance reading (close to zero ohms) indicates continuity, while a high resistance reading indicates an open circuit.

Voltage Divider Analysis:

Resistance measurement is crucial for analyzing voltage divider circuits, where resistors divide the voltage across a circuit. Measuring resistor values helps determine voltage distribution and circuit behavior.

By employing accurate measurement techniques and understanding the applications of resistance measurement in circuit troubleshooting, users can effectively diagnose faults, verify component values, and ensure proper functionality of electrical systems and devices.

CHAPTER NINE

Continuity and Diode Testing

Continuity and diode testing are essential functions of a digital multimeter (DMM) used for circuit diagnostics and component verification. Explanation covering the principles behind continuity testing, procedures for diode testing, and interpretation of test results.

Explanation of Continuity Testing:

Continuity testing is used to determine if there is a continuous electrical connection between two points in a circuit. It helps identify breaks or open circuits, which could prevent proper electrical flow.

The principle behind continuity testing involves sending a small current through the circuit and detecting whether the current flows uninterrupted between the test points.

Procedures for Diode Testing

Setting Up for Diode Testing:

Set the digital multimeter to the diode test mode using the selector dial. The diode test mode is often denoted by a diode symbol or the letters "DIODE."

Ensure the circuit or component under test is disconnected from any power source.

Diode Polarity Identification:

Identify the polarity of the diode to be tested.

Diodes are polarized components, meaning they allow current flow in one direction and block it in the other. The cathode (negative) side is typically marked with a band or line on the diode body.

Diode Testing Procedure:

Place the black (negative) test lead on the cathode side of the diode and the red (positive) test lead on the anode side.

Read the displayed value on the multimeter. A forward-biased diode should show a low voltage drop (typically around 0.5 to 0.7 volts) on the multimeter display. This indicates that current can flow through the diode in the forward direction.

Reverse the polarity of the test leads (black on anode, red on cathode) and repeat the test.

In this configuration, a healthy diode should display "OL" (open circuit) or a high voltage reading, indicating that the diode is blocking current flow in the reverse direction.

Interpretation of Continuity and Diode Test Results

Continuity Test Results:

If the multimeter emits an audible beep or displays a low resistance value (typically close to zero ohms), it indicates continuity, meaning there is a continuous electrical path between the test points.

A high resistance reading or no beep indicates an open circuit, suggesting a break in the circuit path.

Diode Test Results

Forward Bias: A low voltage reading (around 0.5 to 0.7 volts) indicates a forward-biased diode, meaning current can flow through the diode in the forward direction.

Reverse Bias: An "OL" (open circuit) or a high voltage reading indicates a reverse-biased diode, meaning the diode is blocking current flow in the reverse direction.

A short circuit or extremely low voltage reading in both polarities may indicate a faulty or shorted diode.

By conducting continuity and diode tests, technicians can quickly identify faults such as open circuits, short circuits, or faulty diodes within electronic circuits.

Understanding the principles behind these tests and accurately interpreting the results are crucial for efficient troubleshooting and repair of electrical systems and devices.

CHAPTER TEN

Advanced Functions of Digital Multimeters

Digital multimeters (DMMs) often offer advanced functions beyond basic voltage, current, and resistance measurements. These additional features include capacitance, frequency, and temperature measurement capabilities. Explanation of these advanced functions and their practical applications.

Overview of Additional Functions

Capacitance Measurement:

Capacitance measurement allows DMMs to determine the capacitance of capacitors in electronic circuits.

To measure capacitance, the DMM charges the capacitor with a known voltage and measures the resulting charge stored in the capacitor. It then calculates the capacitance value using the formula $C = Q/V$, where C is capacitance, Q is charge, and V is voltage.

Capacitance measurement is useful for verifying capacitor values, diagnosing capacitor health, and selecting appropriate capacitors for circuit design or repair.

Frequency Measurement:

Frequency measurement enables DMMs to determine the frequency of periodic waveforms, such as AC signals.

DMMs typically measure frequency by counting the number of waveform cycles within a specified time period.

Frequency measurement is essential for troubleshooting circuits involving oscillators, generators, motors, and other frequency-dependent components.

Temperature Measurement:

Some advanced DMM models feature temperature measurement capabilities using thermocouples or built-in temperature sensors.

Temperature measurement is crucial for monitoring temperature variations in electrical components, circuits, and environmental conditions.

It enables technicians to identify overheating issues, monitor thermal performance, and ensure safe operation of electronic devices.

Practical Applications of Advanced Functions

Troubleshooting Capacitor Circuits:

Capacitance measurement helps diagnose faulty or degraded capacitors in electronic circuits. A significant deviation from the expected capacitance value may indicate a defective or aged capacitor requiring replacement.

In circuit design and repair, capacitance measurement ensures the selection of capacitors with the correct values to meet circuit specifications.

Frequency Analysis in Power Systems:

Frequency measurement is essential for analyzing power system stability and performance. Variations in frequency can indicate load fluctuations, generator speed deviations, or disturbances in the power grid.

DMMs equipped with frequency measurement capabilities assist power engineers in monitoring grid frequency, diagnosing power quality issues, and optimizing system reliability.

Temperature Monitoring in HVAC Systems:

DMMs with temperature measurement capabilities are valuable tools for HVAC (Heating, Ventilation, and Air Conditioning) technicians.

Temperature measurement allows technicians to assess heating and cooling system performance, detect temperature anomalies in HVAC components, and ensure indoor comfort and energy efficiency.

Electronic Design and Prototyping:

Advanced DMMs play a crucial role in electronic design and prototyping by providing accurate measurements of capacitance, frequency, and temperature.

Engineers and hobbyists use these measurements to validate circuit designs, characterize electronic components, and optimize system performance before final implementation.

By leveraging the advanced functions of digital multimeters, technicians, engineers, and hobbyists can conduct comprehensive electrical measurements, troubleshoot complex circuits, and optimize system performance in various applications across industries. These advanced capabilities enhance efficiency, accuracy, and reliability in electronic testing and diagnostics.

CHAPTER ELEVEN

Data Logging and Recording

Data logging capabilities in digital multimeters (DMMs) enable users to capture and store measurement data over time, providing valuable insights into the behavior of electrical systems and components. Explanations covering the introduction to data logging capabilities, procedures for recording measurements over time, and the importance of data logging in electrical troubleshooting and analysis.

Introduction to Data Logging Capabilities

Digital multimeters equipped with data logging capabilities allow users to record and store measurement data at regular intervals over a specified period.

These DMMs typically feature internal memory or external storage options such as SD cards or USB drives to save the logged data. Data logging functions may include.

Sampling Rate: Users can set the frequency at which measurements are taken, ranging from milliseconds to hours, depending on the application requirements.

Logging Duration: Users can specify the duration of the data logging session, allowing for short-term or long-term monitoring of electrical parameters.

Data Storage: Measurement data is stored in digital format, making it easy to retrieve, analyze, and export for further processing.

Display Options: Some DMMs offer real-time display of logged data, allowing users to visualize trends and patterns directly on the multimeter screen.

Procedures for Recording Measurements Over Time

Setup Parameters:

Set up the digital multimeter for the desired measurement parameters, including the measurement function (voltage, current, resistance, etc.), range, and any additional settings such as sampling rate and logging duration.

Initiate Data Logging:

Start the data logging process using the designated function or button on the DMM.

Ensure that the multimeter is securely positioned and connected to the circuit or device under test.

Monitor and Record Data:

As the data logging session progresses, the multimeter continuously records measurement data at the specified intervals. Monitor the process to ensure proper operation and data integrity.

Review Logged Data:

Once the data logging session is complete, review the logged data either directly on the multimeter display or by transferring it to a computer for analysis. Data can typically be exported in various file formats for further processing.

Importance of Data Logging in Electrical Troubleshooting and Analysis

Fault Diagnosis: Data logging enables technicians to capture intermittent faults or anomalies that may occur over time, aiding in the diagnosis of complex electrical issues.

Performance Monitoring: Continuous monitoring of electrical parameters such as voltage, current, and temperature allows for the assessment of system performance and the detection of trends or deviations from normal operation.

Load Profiling: Data logging facilitates load profiling by recording electrical consumption patterns over time, helping identify peak usage periods and optimize energy efficiency.

Predictive Maintenance: Trend analysis of logged data can help predict equipment failures or performance degradation, allowing for proactive maintenance to prevent downtime and costly repairs.

Regulatory Compliance: In some industries, data logging may be required to comply with regulatory standards or documentation requirements for quality assurance and safety.

Leveraging data logging capabilities in digital multimeters, users can gain valuable insights into the behavior of electrical systems, improve troubleshooting efficiency, optimize system performance, and ensure compliance with industry standards and regulations.

CHAPTER TWELVE

Importance in Electrical Work

Digital multimeters (DMMs) play a crucial role in electrical maintenance, troubleshooting, and installation across a wide range of industries. Explanation covering the importance of digital multimeters in electrical work, their impact on safety and efficiency, and their applications in various industries.

Role of Digital Multimeters in Electrical Maintenance and Troubleshooting

Measurement Versatility: Digital multimeters are versatile tools capable of measuring various electrical parameters such as voltage, current, resistance, capacitance, frequency, and temperature.

This versatility allows technicians to perform comprehensive diagnostics and troubleshooting of electrical systems.

Fault Detection: DMMs help detect and diagnose electrical faults, including open circuits, short circuits, voltage irregularities, and component failures. By accurately measuring parameters, technicians can pinpoint the source of problems and implement effective solutions.

Continuity Testing: Digital multimeters enable continuity testing to verify the integrity of electrical connections and identify open or short circuits. This helps ensure proper electrical continuity and reliability in circuits and systems.

Data Logging and Analysis:
Advanced DMMs with data logging capabilities allow technicians to record and analyze measurement data over time. This facilitates trend analysis, performance monitoring, and predictive maintenance, leading to enhanced system reliability and efficiency.

Impact on Safety and Efficiency in Electrical Installations

Safety Assurance: Digital multimeters enhance safety in electrical installations by providing accurate measurements and identifying potential hazards. By detecting abnormal voltages, currents, or temperatures, DMMs help prevent electrical accidents and ensure compliance with safety standards.

Efficiency Improvement: DMMs streamline electrical troubleshooting and maintenance tasks, reducing downtime and minimizing disruptions to operations. With their quick and accurate measurements, technicians can efficiently identify and resolve issues, leading to improved productivity and cost savings.

Quality Assurance: By ensuring precise measurements and reliable performance, digital multimeters contribute to the quality assurance of electrical installations. Proper measurement and verification of electrical parameters help maintain compliance with regulatory requirements and industry standards.

Applications in Various Industries

Electronics Industry: In the electronics industry, DMMs are used for testing and quality control of electronic components, circuit boards, and devices. They facilitate component characterization, circuit analysis, and functional testing in electronic manufacturing processes.

Automotive Industry: Digital multimeters are essential tools for diagnosing electrical problems in automotive systems, including engine management, ignition, and vehicle electronics. They help automotive technicians troubleshoot electrical issues and maintain vehicle performance and safety.

Aerospace Industry: In the aerospace sector, DMMs are utilized for testing and maintenance of aircraft electrical systems, avionics, and instrumentation. They play a critical role in ensuring the reliability and safety of aircraft systems and components.

Industrial Maintenance: Digital multimeters are widely used in industrial settings for routine maintenance, troubleshooting, and monitoring of electrical equipment and machinery. They help ensure operational efficiency, prevent downtime, and maintain workplace safety.

Overall, digital multimeters are indispensable tools in electrical work, offering precision, versatility, and reliability for maintenance, troubleshooting, and installation tasks across diverse industries. Their impact on safety, efficiency, and quality makes them essential instruments for electrical professionals worldwide.

CHAPTER THIRTEEN

Calibration and Maintenance of Digital Multimeters

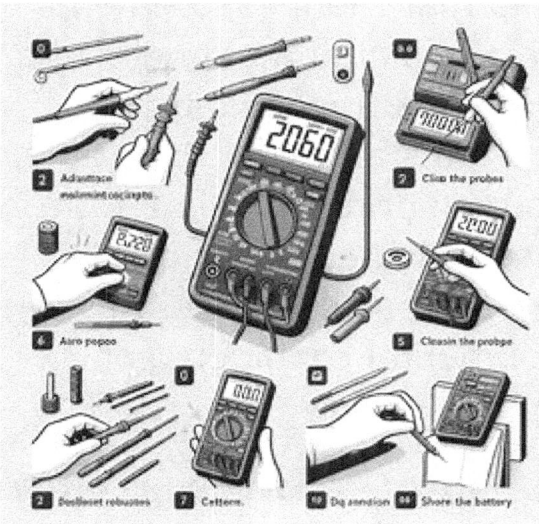

Calibration and maintenance are essential aspects of ensuring the accuracy, reliability, and longevity of digital multimeters (DMMs). Explanation covering the importance of calibration, procedures for calibrating a DMM, and routine maintenance practices.

Importance of Calibration for Accuracy

Accuracy Assurance: Calibration ensures that the measurements taken by a digital multimeter are accurate and traceable to recognized standards. Over time, DMMs may drift or deviate from their specified accuracy due to factors such as component aging, environmental conditions, or mechanical stress.

Compliance: Calibration is often a requirement for regulatory compliance, quality assurance, and adherence to industry standards. Many industries, including aerospace, automotive, electronics, and manufacturing, mandate regular calibration of measurement instruments to ensure product quality, safety, and regulatory compliance.

Reliability: Accurate measurements are crucial for diagnosing faults, troubleshooting electrical systems, and ensuring the safety and reliability of electrical installations. Proper calibration helps maintain the reliability of DMMs, reducing the risk of errors or inaccuracies in critical measurements.

Procedures for Calibrating a Digital Multimeter

External Calibration: External calibration involves comparing the readings of the DMM against a calibrated reference standard, such as a precision resistor or voltage source. The process typically requires specialized calibration equipment and procedures performed by trained technicians or accredited calibration laboratories.

Internal Calibration: Some digital multimeters feature internal calibration routines or self-calibration capabilities. These routines allow the DMM to adjust its internal circuitry and compensate for factors affecting measurement accuracy, such as temperature variations or component drift. Internal calibration procedures are often initiated through the multimeter's menu system or user interface.

Periodic Calibration: Calibration should be performed periodically at regular intervals according to the manufacturer's recommendations, industry standards, or regulatory requirements. The frequency of calibration depends on factors such as the DMM's usage, environmental conditions, and the criticality of measurements.

Routine Maintenance Practices

Visual Inspection: Regularly inspect the digital multimeter for signs of physical damage, wear, or contamination. Check the condition of the test leads, connectors, switches, and display screen. Clean the multimeter's exterior with a soft, dry cloth as needed.

Test Lead Verification: Verify the integrity and accuracy of the test leads by performing continuity checks and resistance measurements using a known reference standard. Replace damaged or worn test leads to maintain measurement accuracy and safety.

Battery Check: Monitor the battery status of the DMM and replace batteries as needed to ensure uninterrupted operation.

Low battery voltage can affect measurement accuracy and stability.

Environmental Control: Store the digital multimeter in a clean, dry environment away from excessive heat, moisture, dust, or corrosive substances. Avoid subjecting the DMM to mechanical shocks or extreme temperature fluctuations that could affect its performance.

Firmware Updates: Periodically check for firmware updates or software patches provided by the manufacturer to address known issues, improve functionality, or add new features to the digital multimeter.

Adhering to proper calibration and maintenance practices, users can ensure the accuracy, reliability, and longevity of digital multimeters, thereby enhancing safety, efficiency, and confidence in electrical measurements and troubleshooting activities.

CHAPTER FOURTEEN

Safety Precautions and Best Practices for Using a Digital Multimeter

Using a digital multimeter (DMM) involves working with electrical circuits and components, which can pose various safety hazards if proper precautions are not followed. Explanation covering an overview of safety hazards, safety precautions, and best practices for handling DMMs.

Overview of Safety Hazards

Electrical Shocks: One of the most significant hazards associated with using a DMM is the risk of electrical shock.

Direct contact with live circuits or accidental contact with energized components can result in electric shock, which can cause serious injury or even death.

Arc Flash: Working on high-voltage circuits or in environments with high fault currents increases the risk of arc flashes. Arc flashes produce intense heat, light, and pressure, posing severe burn and blast hazards to personnel nearby.

Damage to Equipment: Incorrect use of a DMM, such as using incorrect settings or exceeding measurement ranges, can lead to damage to the multimeter itself or the circuit under test. This can result in costly repairs or replacements.

Safety Precautions to Avoid Electrical Shocks and Damage

Personal Protective Equipment (PPE):

Wear appropriate PPE, including insulated gloves, safety goggles, and non-conductive footwear, when working with electrical circuits to protect against electrical shocks and arc flashes.

Verify Circuit Conditions:

Before making measurements, ensure that circuits are de-energized and properly isolated. Use a voltage tester or proximity detector to confirm that circuits are safe to work on.

Use Proper Rating and Range:

Select the correct measurement function, range, and probe configuration on the DMM for the intended application. Avoid exceeding the maximum voltage, current, or resistance ratings of the multimeter.

Beware of High Energy Sources

Exercise caution when working with high-voltage or high-current circuits. Use appropriate current shunts, voltage dividers, or isolation barriers to protect the multimeter and personnel from overloads.

Keep Hands Dry and Free:

Ensure hands are dry and free from moisture or conductive materials when handling a DMM or making measurements to prevent inadvertent short circuits or electrical shocks.

Best Practices for Handling, Storing, and Transporting a Digital Multimeter

Secure Storage:

Store the digital multimeter in a protective carrying case or pouch when not in use to prevent damage from dust, moisture, or accidental impact.

Proper Handling:

Handle the DMM with care, avoiding rough handling or dropping that could damage internal components or affect measurement accuracy.

Transportation Safety

When transporting a DMM, ensure it is securely stored in its carrying case or a padded container to prevent damage during transit. Avoid subjecting the multimeter to extreme temperatures or mechanical shocks.

Regular Inspection:

Periodically inspect the DMM for signs of wear, damage, or malfunction. Check the integrity of test leads, connectors, probes, and the display screen.

Perform routine maintenance and calibration as necessary to ensure reliable performance.

User Manual Reference:

Refer to the manufacturer's user manual for specific safety guidelines, operating instructions, and maintenance procedures for the digital multimeter model being used.

Following these safety precautions and best practices, users can mitigate risks associated with using a digital multimeter, protect personnel and equipment from harm, and ensure safe and effective electrical measurements and troubleshooting activities.

www.ingramcontent.com/pod-product-compliance
Lightning Source LLC
Chambersburg PA
CBHW070349230526
45471CB00006B/2485